T0225206

Cambridge Elements ≡

Elements of Paleontology

BEYOND HANDS ON

Incorporating Kinesthetic Learning in an Undergraduate Paleontology Class

David W. Goldsmith
Westminster College

Paleontological
S O C I E T Y

CAMBRIDGE
UNIVERSITY PRESS

CAMBRIDGE
UNIVERSITY PRESS

University Printing House, Cambridge CB2 8BS, United Kingdom

One Liberty Plaza, 20th Floor, New York, NY 10006, USA

477 Williamstown Road, Port Melbourne, VIC 3207, Australia

314–321, 3rd Floor, Plot 3, Splendor Forum, Jasola District Centre,
New Delhi – 110025, India

79 Anson Road, #06–04/06, Singapore 079906

Cambridge University Press is part of the University of Cambridge.

It furthers the University's mission by disseminating knowledge in the pursuit of
education, learning, and research at the highest international levels of excellence.

www.cambridge.org
Information on this title: www.cambridge.org/9781108717878
DOI: 10.1017/9781108681544

First published 2018

A catalogue record for this publication is available from the British Library.

ISBN 978-1-108-71787-8 Paperback
ISSN 2517-780X (online)
ISSN 2517-7796 (print)

Beyond Hands On

Incorporating Kinesthetic Learning in an Undergraduate Paleontology Class

Elements of Paleontology

DOI: 10.1017/9781108681544
First published online: October 2018

David W. Goldsmith
Westminster College

Abstract: Hands-on learning in paleontology, and geology in general, is fairly common practice. Students regularly use rocks, fossils, and data in the classroom throughout their undergraduate career, but they typically do it sitting in a chair in a lab. Kinesthetic learning is a teaching model that requires students to be physically active while learning. Students may be involved in a physical activity during class or might be using their own bodies to model some important concept. This Element briefly discusses the theory behind kinesthetic learning and how it fits into a student-centered, active learning classroom. It then describes in detail methods for incorporating kinesthetic learning into student exercises on biostratigraphy, assessment of sampling completeness, and modeling evolutionary processes. Assessment data demonstrate that these exercises have led to significantly improved student learning outcomes tied to these concepts.

Keywords: Learning Styles, Geoscience Education, Assessment

ISBNs: 9781108717878 (PB), 9781108681544 (OC)
ISSNs: 2517-780X (online), 2517–7796 (print)

Contents

1 Introduction

Since the 1970s, science education in America has moved increasingly away from a traditional, teaching-centered paradigm and has instead moved steadily toward a student-centered paradigm (Clasen and Bowman, 1974; Volpe, 1984; Felder and Brent, 1996; Michael, 2006). A teaching-centered paradigm regards the educational process as an exchange of factual knowledge, with the student's role relegated to that of passive note-taker (Volpe, 1984). Conversely a student-centered paradigm regards the educational process as an opportunity for students to understand the foundations of that knowledge. The student-centered paradigm incorporates active learning strategies and an awareness of student learning styles to help students discover underlying concepts and become active and engaged participants in their own education (Weimer, 2002; Brown Wright, 2011).

One of the best predictors of student success in science, technology, engineering, and mathematics (STEM) classes is student engagement. Increased engagement leads to better comprehension of course material and concepts (Prince, 2004). This pattern has been particularly well demonstrated in physics education research through pre and post testing using the Force Concept Inventory (see Hestenes et al., 1992; Hake, 1998; Hollewarth and Moelter, 2011). These studies particularly acknowledge active learning as a powerful mechanism for improving engagement and therefore learning. Students in active learning classrooms do not just sit passively and receive knowledge. They discuss, write, and think during class to a greater extent than ever before (Michael and Modell, 2003), and therefore engage with the material to a deeper degree than in traditional lecture classes (Chi and Wylie, 2014). These active learning strategies increase not only comprehension but also retention in STEM programs. In 2012, the President's Council of Advisors on Science and Technology (2012) cited an increase in active learning strategies as a key to improving not just the quality of STEM education but also STEM retention.

Part of the movement toward more active classroom environments has been a push for more hands-on learning, particularly in STEM fields (Flick, 1993; Carlson and Sullivan, 1999; Stohlmann et al., 2012; Christensen et al., 2015). STEM students typically engage in hands-on learning through lab exercises, either separate from or integrated into class sessions (Hofstein and Lunetta, 2004). However, there are ways to engage students' physical learning style beyond hands-on manipulation of lab materials. Kinesthetic learning requires students to use their entire bodies in the learning process and thus taps into an additional learning modality beyond the hands-on.

One of the most widely known classifications of learning styles is Fleming's VARK model, which separates students into Visual, Auditory, Read/Write, and Kinesthetic learners (Fleming 1995). There is ample evidence to support the notion that different students learn in different ways, and furthermore that very few people learn effectively across all learning modes (Fleming, 1995; Anbarasi et al., 2015). Unfortunately, many people take the wrong lesson from these studies and assume that teaching styles must precisely match learning styles to achieve optimal results. Most teachers have had a student inform them, "I learn better when I see things written down," or "I'm more of a visual learner." Likewise, many educators have to work to identify individual students' learning styles in order to adapt their teaching style to meet each student's needs (see, for example, Bull and McCalla, 2002, or Hawk and Shah, 2007). This practice of diagnosing a student's particular learning style and then tailoring instruction to best fit that learning style is often called "meshing." It is the logical extension of identifying students' learning styles, but, according to most research on the topic, it is misguided (Kang Sheng, 2016). Students, as it turns out, do not learn best when teaching methods are matched to their learning styles, but rather when they are given the opportunity to learn across a variety of learning styles (Griggs et al., 2009). Even the most specialized learner benefits from the opportunity to learn through multiple modalities (Lujan and DiCarlo, 2006).

Traditional undergraduate classrooms provide many opportunities for auditory and visual learning. Lectures and PowerPoint presentations are the standards for those modalities. As more and more colleges embrace the concept of "writing across the curriculum," more read/write-centered learning is happening in science classes. And of course, lab sessions are typically the place where hands-on learning happens. But hands-on learning is not perfectly synonymous with kinesthetic learning. More recent classifications of learning styles have separated hands-on from kinesthetic learning (Favre, 2009).

2 What Is Kinesthetic Learning?

Kinesthetic learning is more than simply hands-on or experiential learning. It was originally described by Dunn and Dunn as a learning style in which students require whole body movement to help them process new information (1978). So, whereas a hands-on exercise would involve students manipulating some model, or actively working with data, in a kinesthetic exercise, the students themselves either complete a particular physical task, or they themselves are the model. Kinesthetic learning has been demonstrably successful in helping students to master concepts in a range of STEM disciplines, including chemistry (Bridgeman et al., 2013; Anbarasi et al., 2015) and mathematics (Beaudoin and Johnston, 2011), and even disciplines outside of STEM such as poetry (Zimmerman, 2002).

As a geological example, students learning the depth of geologic time through a hands-on learning model will very frequently build some model of the timescale proportioned to some object (deep time as a clock, a calendar, etc.). The punchline from such an exercise is typically something like, "If time were a calendar, humans only evolve on December 31, and the dinosaurs went extinct on Christmas Eve." The problem with these exercises is that anything a student can model in a classroom is too small to adequately express the depth of geologic time. Even students who scale geologic time to some large skyscraper like Burj Khalifa or the diameter of the planet use maps and models. On these scales, order-of-magnitude differences can still look negligible.

Students learning the same concept kinesthetically might actually pace off a geologic timescale. A starting question might be, "If one step represented 1 million years (the approximate life span of the human species), how far would you need to go to walk back in time to the extinction of the dinosaurs?" Assuming an average stride length of three feet, students quickly find themselves nearly 200 feet from their starting point to get to the Mesozoic; and, if they were to continue, they would need to hike nearly an eighth of a mile before they encountered a trilobite. This kinesthetic exercise engages students physically as well as mentally, and has the added benefit of scaling up the typical classroom exercise to a point where it drives home the learning goals more concretely.

2.1 Degrees of Kinesthetic Learning in Paleontology

There are many different degrees of kinesthetic learning. As previously discussed, some authors would consider any kind of lab work kinesthetic as long as students are actually working with something physical (Mobley and Fisher, 2014). For other authors (Favre, 2009), kinesthetic learning requires that the students use their own bodies in some meaningful way during the learning process. This can mean simply that students are moving while learning, or, in the purest form of kinesthetic learning, that the students model some concept or process using their own bodies as part of the model.

In geology, the most common form of kinesthetic learning is a field trip. However, field trips can provide logistical challenges. Field trips create potential problems with liability, they require obtaining permits or permission for land use, and they need vehicles and people to drive them. Field trips can be particularly challenging for students with mobility difficulties or visual impairments. Furthermore, field trips are time-consuming and cannot practically be used to illustrate every concept in paleontology. Kinesthetic learning activities can be simple, engaging, and completed within a typical class time. Here I describe several strategies for incorporating kinesthetic learning methods into traditional paleontology lab activities, two in which students move while learning, and one in which students not only move, but use their bodies as part of a physical model.

All of the exercises described here are used in the same class. Principles of Paleontology is an upper-division course intended for students in their junior year of college. It is a required course for geology majors at Westminster College and an elective option for geology minors as well as for students in biology and environmental studies. There are no prerequisites for the course, but students are strongly encouraged to have taken an introductory geology course and an introductory organismal biology course before enrolling. It is offered once each academic year, in the fall semester, and has a typical enrollment of 15–24 students. Of these students, half are typically geology majors and half are from other disciplines. The course meets twice a week for a two-hour session that integrates lecture, lab activities, and journal discussions.

3 Moving While Learning

3.1 Moving While Learning 1: Fossil Correlation

Textbooks and the Internet are filled with examples of fossil correlation exercises, many of which fit a general pattern. A typical exercise might include diagrams of multiple cliff faces, each divided into several strata. Each stratum then contains multiple fossils. Students are required to align strata with similar fossils to work out age relations between the strata and create a complete stratigraphic column for the region. Bennington (2018) provides a particularly good example of this style of exercise.

These exercises are effective, but oversimplified. By giving students the data to work with in a series of tables or diagrams, they omit the data-gathering process. Furthermore, they can create the false impression that paleontologists are always able to see all the layers of a formation simultaneously. In reality, paleontologists more frequently need to integrate several discrete observations within their field area to get a true sense of age relations in the region and then correlate them with other exposures miles or tens of miles away.

In the kinesthetic correlation exercise I use in class, students gather data in discrete bits and must integrate them before they can do their fossil correlation. The exercise is spread across four different multistory buildings on campus. I print out 17 different pictures of fossil assemblages and place one in each floor of the stairwell of each building. I use pictures of fossils rather than actual specimens because the time required to complete the exercise typically exceeds the allotted class time and so the specimens must stay in place in public access buildings overnight as students complete the exercise for homework. I put assemblages in stairwells so that once students find one assemblage, they can easily find the rest by moving purely vertically through the building. To provide access for students with mobility issues, I also make sure that each floor's fossil assemblage can be reached from the buildings' elevators. Students are required to move around campus, find the fossils in each building, identify them using a provided identification key, and record their positions in each building relative to one another. In addition to the fossil horizons, one floor of one building is labeled "Unconformity: An unknown amount of time is missing." Figure 1 shows the distribution of fossil taxa among buildings that students record as their data.

Once students have collected their data, they are responsible for putting the fossil taxa in their proper temporal order. Students may use Steno's Law of Superposition to determine the age relations of fossils within each building, but they may not assume lateral continuity between buildings. In order to determine the age relations between the "strata" in different buildings, students must rely on fossil correlation. As often happens with actual fossil occurrences, very few taxa occur in only one stratum, and very few strata have only one taxon. As a result, students must rely on combinations of taxa to make their assessments. Figure 2 shows how students can use their data to put the strata in temporal order.

As a final step in this exercise, and in order to link principles of stratigraphy back into other topics in paleontology, students must use their completed sequence of strata to choose between two different possible cladograms for a subset of the taxa in the exercise. One cladogram has a branching order largely consistent with the order in which the fossils occur in the stratigraphic column. The other has late-appearing taxa branching very early. This final step introduces students to the concept (though not the term) of *stratigraphic debt* (Smith, 1994) and leads into the next day's discussion on integrating the fossil record into evolutionary hypotheses.

Converse Hall
Nevadella   *Bradyfallotaspis*
Esmereldina   *Bradyfallotaspis*
Geraldinella   *Esmereldina*   *Bradyfallotaspis*
Fallotaspis   *Parafallotaspis*
Fallotaspis   *Daguinaspsis*

Foster Hall
Olenellus
Olenellus
Nevadella   *Bradyfallotaspis*
Unconformity
Fallotaspis   *Daguinaspsis*
Fallotaspis
Profallotaspis   *Eofallotaspis*

Meldrum   Science Center
Fallotaspis
Profallotaspis   *Eofallotaspis*
Profallotaspis

Gore Building
Olenellus
Nevadella   *Bradyfallotaspis*
Nevadella   *Bradyfallotaspis*

Figure 1 Distribution of imaginary fossil horizons across four different buildings on the Westminster College campus. Horizons are placed in or near stairwells so that they are as close as possible to one another in a vertical column.

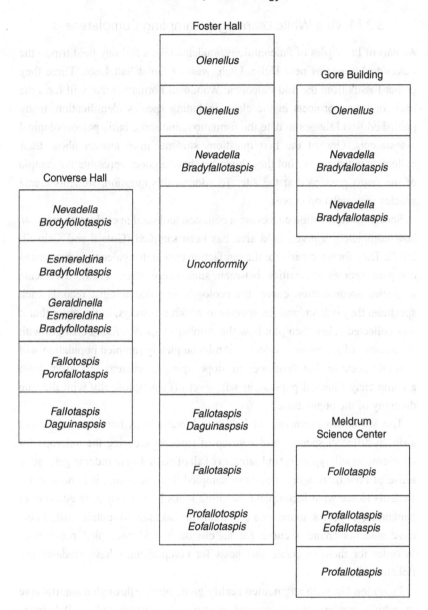

Figure 2 Combined stratigraphic column showing the proper age relations of all fossils within the exercise. The unconformity in Foster Hall actually represents a considerable period of missing time.

3.2 Moving While Learning 2: Sampling Completeness

As part of Principles of Paleontology, students take a full-day field trip to the Lakeside Mountains near Delle, Utah, west of Great Salt Lake. There they gather fossils from the mid-Paleozoic Woodman Formation that will form the basis of future projects in the class, including species identification, using published fossil ranges to date the formation, and some basic paleoecological assessments. One of the first questions students must answer about their collection is whether or not they think they have a good representative sample of the fauna preserved at the site. To address this question, students create species accumulation curves.

Species accumulation curves are a common tool used in ecology to estimate how completely a given field area has been sampled (Gotelli and Calwell, 2011). They form the basis for the rarefaction curves that paleontologists use to compare species diversities between sites (Sorgenfrei, 1958). To create a species accumulation curve, the ecologist or paleontologist records each specimen they collect, and the species to which it belongs, in the order that it was collected. They then plot how the number of species found changes with the number of specimens collected. An incompletely sampled population will generate a curve that continues to slope upward, whereas the curve from a completely sampled population will level off asymptotically with the true diversity of the population.

The Woodman Formation at Lakeside Mountains is home to only about a dozen different species, and a group of students scouring the outcrops for an afternoon will typically find samples of all of them. But in order to get a good sense of how thoroughly they have sampled their field site, it is helpful for students to see what incomplete sampling looks like. As such, in addition to generating a species abundance curve for the Lakeside Mountains site, I also have students create a curve for an obviously under-sampled population. In order for them to create this basis for comparison, I have students use Pokémon Go.

Pokémon Go is an augmented reality game played through a smartphone in which players travel around capturing monsters called Pokémon. Pokémon come in hundreds of different types, each easily distinguishable from one another. Like many college campuses, Westminster College is home to many Pokémon gyms, lures, and other places where Pokémon can typically be found. As a result, students can easily collect several Pokémon in just a few hours.

Before our session discussing sampling completeness, students download the Pokémon Go app to their phones and spend a few hours of the weekend recording what Pokémon they encounter. Students share a list of their encounters to an online spreadsheet. They then have a two-part assignment to complete for class. In the first part of their assignment, students make a simple graph showing how the number of different Pokémon types they encountered increased with the number of individuals they met.

For the second part of the assignment, students must take the specimen counts from the Lakeside Mountains collection and find a way to create a second curve showing how diversity increases with sample size. For this part of the assignment, students have an extra challenge. Students, working in separate teams of two over the course of the day, collect fossils from the Lakeside Mountains, and do not identify the species they have found until the next week's lab session. This makes it difficult to keep track of exactly which fossils were found in which order. Students have a list of how many specimens were found for each species, but not the same chronological data that they had in the Pokémon exercise to make an accumulation curve.

Because students are only working with count data and do not have a list of the order in which the specimens were found, they need to find a way to randomly sample the data without replacement. Because students come into the class with a variety of coding backgrounds, they use several different methods to sample the list. Some write R scripts. Some use Excel. Some even just draw slips of paper with species names out of a hat. Anything that achieves the goal of random sampling will allow them to make the necessary accumulation curve. The two different ways in which students create their Pokémon and Lakeside Mountains accumulation curves, with and without the order in which the specimens were collected, give students a sense of how they might work with their own field data, as opposed to data collected from some online data repository like the Paleobiology Database.

Figure 3 shows typical curves generated by students in the class. Despite the fact that the Pokémon Go motto is "Gotta catch 'em all," students never do. At last count, there are more than 800 different Pokémon types in the game and students typically only sample 100–200 individuals. As a result, the curve of specimens versus species never really levels off. However, a comparable sample size of fossils from the Woodman Formation does level off very quickly. A comparison of these two graphs allows students to see the difference between adequate and inadequate sampling of diversity and to have confidence that their Lakeside Mountains collection is complete enough to represent the actual diversity of the area.

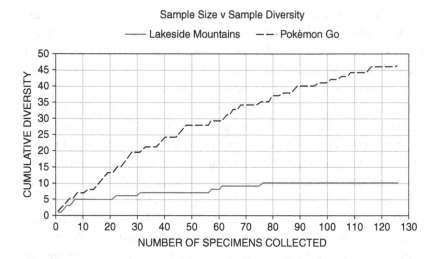

Figure 3 Species accumulation curves showing how sample diversity changes with sample size for two different collections. The dashed line represents a collection of 126 different Pokémon monsters found across the Westminster College campus. The solid line represents a collection of 126 different fossil specimens from the Woodman Formation, Lakeside Mountains, Utah. The Pokémon collection continues to slope upward with increased sampling, indicating that the sample does not yet adequately represent the true diversity of the population. The Woodman Formation line tapers off asymptotically, indicating that the collected specimens are representative of the true diversity of the assemblage.

3.3 Learning by Moving: Literal Random Walks

Random walks are used as a null hypothesis in several evolutionary models, including studies of evolutionary rates and trends (see, for example, Bookstein, 1987; Roopnarine et al., 1999; Hunt, 2006). Unfortunately, despite their broad use in paleontology, ecology, and evolutionary biology, geology majors first setting foot in a paleontology class may be largely unfamiliar with them. Principles of Paleontology introduces students to random walks and their use in the analysis of evolutionary trends, by having students walk almost literally at random.

This exercise takes place on a basketball court in the college gymnasium. The basketball court provides a convenient place to do this exercise because its many different paint lines can serve as points of reference. We begin the exercise with the students lined up, one behind the other, under one basket, out of bounds of the court. Each student is given a homemade random number generator (RNG); that is, a small, clear food storage container with a die inside it. For our first trial, students are given the following instructions:

- One at a time, step onto the court.
- Walk across the court to the other side, rolling a new number on your RNG with each step.
- If the number you roll is odd, then in addition to taking a step forward, take a step to the right.
- If the number you roll is even, then in addition to taking a step forward, take a step to the left.
- When you get to the other end of the court, stop.
- If, when you get to the other end, there is already a person standing where you should be standing, then line up behind them.

Before students take this first random walk, we have a brief discussion about their expectations so that we can see how those expectations match up with reality later on.

Running this first model multiple times yields similar results each time. Students end up arranged in a rough bell curve on the opposite end of the basketball court. The average position of a student relative to the baseline is approximately under that end's basket, exactly where they started from on the opposite end. Students can change the standard deviation of that bell curve by making a simple change to the rules. Instead of left for even and right for odd, students perform the walk again, but this time only moving left if they roll a one and only moving right if they roll a six. With this modification, students still end up in a rough bell curve, but a much higher and narrower one.

Once students understand the basic physical mechanics of this exercise, they can then simulate different evolutionary models just by making small changes to the rules. Table 1 gives examples of how to change the rules to simulate some well-known models. For each of these models, we begin by discussing what students expect the final result to look like and then end by comparing our results as a class to those expectations. We also take a few minutes after each run to think about whether the distributions we have arrived at could have originated through alternative rules.

Figure 4 shows a schematic view of how students end up arranged after a series of different trials. In the first trial (Figure 4a), students simulate a simple random walk and generate a rough bell curve. In the second trial (Figure 4b), students simulate a random walk with a systematic bias. Students step left only on a roll of one or two and to the right on a three or higher. As a result, the probability of a step to the right is 66.7% for each step

Table 1 Instructions necessary to simulate different evolutionary models in a random walk exercise

Model	Rules
Absorbing Boundary aka Gambler's Ruin	• Start three paces to the left of the court's end line. • Proceed as usual. • If at any point a step to the right would make you touch the end line, step out of bounds and stop.
Reflecting Boundary aka Passive Trend	• Start three paces to the left of the court's end line. • Proceed as usual. • If at any point a step to the right would make you touch the end line, step left instead.
Directional Selection aka Driven Trend	• Start three paces to the left of the court's end line. • Proceed as usual, but move to the right on any roll of three or higher. Only step left if you roll a one or two. • If at any point a step to the right would make you touch the end line, step left instead.

forward instead of 50%. In this case, the entire bell curve is shifted to the right and even the leftmost tail of the curve (the student who rolled the most ones and twos) ends up to the right of their starting point. In the third trial (Figure 4c), students simulate an absorbing boundary. They start close to one sideline of the court and follow the same instructions as the first walk, but if at any point they step out of bounds, they stay out of bounds and stop their walk. In this case, only a few students make it to the other end of the court, and those who do still wind up very close to the sideline. This is similar to the Gambler's Ruin model of extinction popularized by Raup (1991). Finally, in the fourth trial (Figure 4d), students simulate a reflecting boundary. Once again, they start close to one sideline and follow the instructions for an unbiased random walk. But this time, if a step would take them out of bounds, they simply take a step forward with no step to the side.

Once students have walked the walk, they go back to the classroom and talk the talk. The classroom discussion for that day looks at several different purported trends in the history of life, including Cope's Rule, that organisms within evolutionary lineages tend to increase in size through time (Cope, 1885), and they increase in organismal complexity through time. Students look at published data sets supporting each trend and discuss which of their random walks the data supporting each trend look most like.

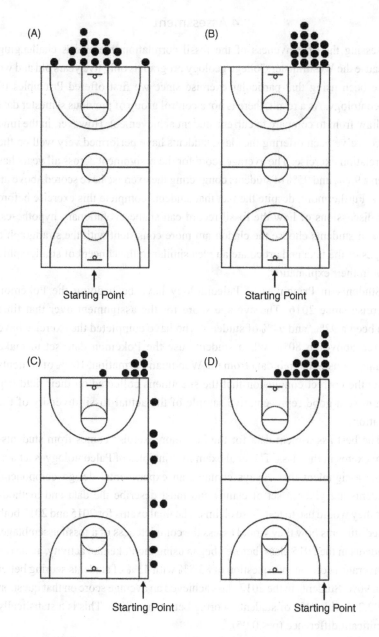

Figure 4 Distribution of students on a basketball court resulting from
different random walk models (A) unbiased random walk (B) biased random
walk (C) random walk with an absorbing boundary (D) random walk with
a reflecting boundary

4 Assessment

Assessing the effectiveness of the fossil correlation exercise is challenging because the Westminster College geology program is only five years old and we have been using that particular exercise since we first offered Principles of Paleontology. As a result, there is not a control group of previous semester data to draw from to compare to current student achievement. However, in the time since we've been offering the class, students have performed very well on the correlation exercise. The average score for the assignment across all years has been a 93%, and 95% of students completing the exercise have scored above an 80%. Furthermore, despite the fact that students complete this exercise before our discussions of how the fossil record can shape evolutionary hypotheses, 90% of students choose the cladogram more consistent with the stratigraphic ranges of the taxa and articulate an idea similar to the concept of stratigraphic debt in their explanation.

Students in Principles of Paleontology have been doing the Pokémon exercise since 2016. The average score for the assignment over that time has been a 92%, and 94% of students who have completed the exercise have scored above an 80%. When students use the Pokémon data set to make comparisons with their data from the Woodman Formation, 100% of students draw the correct conclusion that the specimens collected on their field trip comprise a good representative sample of the actual fossil diversity of the location.

The best assessment data for the Pokémon exercise comes from students' final exams in the class. The final exam in Principles of Paleontology is a take-home assignment that always contains an experimental design component. Students are given a set of claims and must describe the data and methods that they would use to test those claims. The final exams for 2015 and 2017 both asked students how they would assess the completeness of a fossil assemblage. Students in the 2015 class, before I began using the Pokémon activity, achieved an average score on that question of 83.8% with 43% of students scoring better than 90%. Students in the 2017 class achieved an average score on that question of 92.2% with 81% of students scoring better than 90%. This is a statistically significant difference ($p < 0.05$).

Students in Principles of Paleontology have been doing the random walk exercise since 2017, and as yet it does not have a specific assignment attached to it. However, several follow up class sessions revisit the concepts of random walks and passive versus driven trends. Students in Principles of Paleontology read original journal articles frequently over the course of the semester. Two of the articles they read are Niklas's work comparing random

Participation: After each case study session, each student will be graded on the following four criteria using the scale below:				
10 (outstanding)	9 (very good)	8 (good)	7 (satisfactory)	6 (poor)
Preparation			The student's comments show that he/she has come ready to discuss the day's readings.	
Analysis			The student's comments go beyond simple summary of the reading and demonstrate the ability to synthesize and evaluate.	
Active Listening			The student is listening to his/her peers in class and responding to what they have to say.	
Insight			The student is using the text and reflection on his/her peer's comments to move the conversation in interesting directions.	

Figure 5 Grading rubric used for in-class discussions of published papers

walks through a theoretical morphospace of plant evolution to actual plant evolution (1994) and McShea's explorations of mechanisms behind evolutionary trends (1994). I have used these case studies in class for several years, but only recently began prefacing our conversations with the random walk exercise.

Students are assessed after each in-class discussion of a journal article using a rubric that scores them on four specific points, Preparation, Analysis, Active Listening, and Insight (see Figure 5). In semesters prior to 2017, student scores on the discussion of the Niklas paper were among the lowest of the semester, with an average of 32.2/40 (81%). Much of the in-class discussion of this article centered around methods and models. Students wanted clarification on exactly what a random walk was, and on what it meant for a system to have a directional bias. Since incorporating the random walk exercise into the course, conversations about this article dwell less on method and concentrate more on meaning, and scores have increased as a result with a 2017 average of 35.4/40 (89%). This is a statistically significant difference ($p < 0.05$).

5 Conclusions

Increasing student engagement typically increases student learning. Kinesthetic learning exercises are demonstrably effective across a range of disciplines, not only because they increase student engagement but also because they allow students to learn through a different and often untapped learning style. This style of learning can be incorporated into an undergraduate paleontology class to help students understand a range of skills and concepts, including fossil correlation, assessment of sampling completeness, and random walk processes.

Beyond the pedagogical or the practical, there is one additional philosophical reason to embrace kinesthetic learning in paleontology. Paleontology is the study of ancient life. Particularly, it is the study of how that life functioned, moved, and interacted with both its environment and with other living things. Incorporating function, motion, and interaction into paleontological pedagogy emphasizes and reinforces what students are there to learn in the first place.

References

Anbarasi, M., Rajkumar, G., Krishnakumar, S., Rajendran, P., Venkatesan, R., Dinesh, T., Mohan, J., and Venkidusamy, S. (2015). Learning style-based teaching harvests a superior comprehension of respiratory physiology. *Advances in Physiology Education*, **39**, 214–217.

Beaudoin, C. R. and Johnston, P. (2011). The impact of purposeful movement in algebra instruction. *Education*, **132**, 82–96.

Bennington, J. B. (2018). Biostratigraphic and lithostratigraphic correlation of sedimentary strata in the Atlantic Coastal Plain. Retrieved February 1, 2018, https://serc.carleton.edu/NAGTWorkshops/paleo/activities/33136.html.

Bookstein, F. L. (1987). Random walk and the existence of evolutionary rates. *Paleobiology*, **13**, 446–464.

Bridgeman, A. J., Schmidt, T. W., and Young, N. J. (2013). Using atomic orbitals and kinesthetic learning to authentically derive molecular stretching vibrations. *Journal of Chemical Education*, **90**, 889–893.

Brown Wright, G. (2011). Student-centered learning in higher education. *International Journal of Teaching and Learning in Higher Education*, **23**, 92–97.

Bull, S. and McCalla, G. (2002). Modelling cognitive style in a peer help network. *Instructional Science*, **30**, 497–528.

Carlson, L. E. and Sullivan, J. F. (1999). Hands-on engineering: Learning by doing in the integrated teaching and learning program. *International Journal of Engineering Education*, **15**, 20–31.

Chi, M. T. and Wylie, R. (2014). The ICAP framework: Linking cognitive engagement to active learning outcomes. *Educational Psychologist*, **49**, 219–243.

Christensen, R., Knezek, G., and Tyler-Wood, T. (2015). Alignment of hands-on STEM engagement activities with positive STEM dispositions in secondary school students. *Journal of Science Education and Technology*, **24**, 898–909.

Clasen, R. E. and Bowman, W. E. (1974). Toward a student-centered learning focus inventory for junior high and middle school teachers. *Journal of Educational Research*, **68**, 9–11.

Cope, E. D. (1885). On the evolution of the vertebrata, progressive and retrogressive. *American Naturalist*, **19**, 140–148.

Dunn, R. S. and Dunn, K. J. (1978). *Teaching Students through Their Individual Learning Styles: A Practical Approach*. Reston, VA: Reston Publishing.

Favre, L. R. (2009). Kinesthetic instructional strategies: Moving at-risk learners to higher levels. *Insights on Learning Disabilities*, **6**, 29–35.

Felder, R. M. and Brent, R. (1996). Navigating the bumpy road to student-centered instruction. *College Teaching*, **44**, 43–47.

Fleming, N. D. (1995). I'm different; not dumb. Modes of presentation (VARK) in the tertiary classroom. In A. Zelmer, ed., Research and Development in Higher Education, Proceedings of the 1995 Annual Conference of the Higher Education and Research Development Society of Australasia (HERDSA), pp. 308–313.

Flick, L. B. (1993). The meanings of hands-on science. *Journal of Science Teacher Education*, **4**, 1–8.

Gotelli, N. J. and Colwell, R. K. (2011). Estimating species richness. In A. E. Magurran and B. J. McGill, eds., *Biological Diversity: Frontiers in Measurement and Assessment*. Oxford: Oxford University Press, pp. 39–54.

Griggs, L., Barney, S., Brown-Sederberg, J., Collins, E., Keith, S., and Ianucci, L. (2009). Varying pedagogy to address student multiple intelligences. *Human Architecture: Journal of the Society of Self-Knowledge*, **7**, 55–60.

Hake, R. R. (1998). Interactive-engagement versus traditional methods: A six-thousand student survey of mechanics test data for introductory physics courses. *American Journal of Physics*, **66**, 64–74.

Hawk, T. F. and Shah, A. J. (2007). Using learning style instruments to enhance student learning. *Decision Sciences Journal of Innovative Education*, **5**, 1–19.

Hestenes, D., Wells, M., and Swackhammer, G. (1992). Force concept inventory. *The Physics Teacher*, **30**, 141–166.

Hoellwarth, C. and Moelter, M. J. (2011). The implications of a robust curriculum in introductory mechanics. *American Journal of Physics*, **79**, 540–545.

Hofstein, A. and Lunetta, V. N. (2004). The laboratory in science education: Foundations for the twenty-first century. *Science Education*, **88**, 28–54.

Hunt, G. (2006). Fitting and comparing models of phyletic evolution: Random walks and beyond. *Paleobiology*, **32**, 578–601.

Keng Sheng, C. (2016). Tailoring teaching instructions according to students' different learning styles: Are we hitting the right button? *Education in Medicine Journal*, **8**, 103–107.

Lujan, H. L. and DiCarlo, S. E. (2006). First-year medical students prefer multiple learning styles. *Advances in Physiology Education*, **30**, 13–16.

McShea, D. W. (1994). Mechanisms of large-scale evolutionary trends. *Evolution*, **48**, 1747–1763.

Michael, J. (2006). Where's the evidence that active learning works? *Advances in Physiology Education*, **30**, 159–167.

Michael, J. and Modell, H. I. (2003). *Active Learning in Secondary and College Science Classrooms: A Working Model for Helping the Learner to Learn*, 1st edn. London: Lawrence Erlbaum Associates.

Mobley, K. and Fisher, S. (2014). Ditching the desks: Kinesthetic learning in college classrooms. *The Social Studies*, **105**, 301–309.

Niklas, K. J. (1994). Morphological evolution through complex domains of fitness. *Proceedings of the National Academy of Sciences of the United States of America*, **91**, 6772–6779.

The President's Council of Advisors on Science and Technology. (2012). Report to the president – Engage to Excel: Producing one million additional college graduates with degrees in science, technology, engineering, and mathematics. Retrieved February 28, 2018, https://obamawhitehouse.archives.gov/sites/default/files/microsites/ostp/pcast-engage-to-excel-final_2-25-12.pdf.

Prince, M. (2004). Does active learning work? A review of the research. *Journal of Engineering Education*, **93**, 223–231.

Raup, D. M. (1991). *Extinction: Bad Genes or Bad Luck?* 1st edn. New York, NY: W. W. Norton & Co.

Roopnarine, P. D., Byars, G., and Fitzgerald, P. (1999). Anagenetic evolution, stratophenetic patterns, and random walk models. *Paleobiology*, **25**, 41–57.

Sorgenfrei, T. (1958). *Molluscan Assemblages from the Marine Middle Miocene of South Jutland and Their Environments*. Copenhagen: Geological Survey of Denmark.

Smith, A. B. (1994). *Systematics and the Fossil Record: Documenting Patterns of Evolution*, 1st edn. Oxford: Blackwell's Science.

Stohlmann, M., Moore, T. J., and Roehrig, G. H. (2012). Considerations for teaching integrated STEM education. *Journal of Pre-College Engineering Education*, **2**, 28–34.

Volpe, E. P. (1984). The shame of science education. *Integrative and Comparative Biology*, **24**, 433–441.

Weimer, M. (2002). *Learner-Centered Teaching: Five Key Changes to Practice*, 1st edn. San Francisco, CA: Jossey-Bass Publishers.

Zimmerman, V. (2002). Moving poems: Kinesthetic learning in the literature classroom. *Pedagogy*, **2**, 409–425.

Cambridge Elements ≡

Elements of Paleontology

Editor-in-Chief
Colin D. Sumrall
University of Tennessee

About the Series
The Elements of Paleontology series is a publishing collaboration between the Paleontological Society and Cambridge University Press. The series covers the full spectrum of topics in paleontology and paleobiology, and related topics in the Earth and life sciences of interest to students and researchers of paleontology.

The Paleontological Society is an international nonprofit organization devoted exclusively to the science of paleontology: invertebrate and vertebrate paleontology, micropaleontology, and paleobotany. The Society's mission is to advance the study of the fossil record through scientific research, education, and advocacy. Its vision is to be a leading global advocate for understanding life's history and evolution. The Society has several membership categories, including regular, amateur/avocational, student, and retired. Members, representing some 40 countries, include professional paleontologists, academicians, science editors, Earth science teachers, museum specialists, undergraduate and graduate students, postdoctoral scholars, and amateur/avocational paleontologists.

Paleontological
S O C I E T Y

Elements of Paleontology

Elements in the Series

These Elements are contributions to the Paleontological Short Course on *Pedagogy and Technology in the Modern Paleontology Classroom* (organized by Phoebe A. Cohen, Rowan Lockwood, and Lisa Boush), convened at the Geological Society of America Annual Meeting in November 2018 (Indianapolis, Indiana USA).

A full series listing is available at: www.cambridge.org/EPLY